BY
FREDERIC T. BIOLETTI

British Library Cataloguing-in-Publication Data
A catalogue record for this book is available from
the British Library

FREDERIC THEODORE BIOLETTI

Frederic Theodore Bioletti was born in 1865 in Liverpool, England.

In 1878, he emigrated to the United States and resided in Sonoma County, California. He attended Heald's Business School in San Francisco before beginning, what would become his life's vocation, working for Senator Stanford in his commercial wine cellar at Vina Ranch.

From 1889 to 1900, Bioletti studied at the University of California, Berkeley, where he received his Bachelor's and Master's degrees in 1894 and 1898 respectively. While there, he was an assistant to Professor E. W. Hilgard, a noted soil scientist, with whom he studied the fermentation of wines. Their work greatly influenced the vintner's of California and resulted in a higher quality grape being produced in the region.

Bioletti left California for South Africa, in 1901, to teach viticulture, oenology, and horticulture, but returned three years later to rejoin the University at Berkeley. For most of the remainder of his career he taught and conducted research at the University's Department of Viticulture and Oenology where he was both their first professor and first chair of the department. He also founded the grape breeding program at the University of California Agricultural Experiment Station where he was active in introducing and breeding new varieties of grape. During prohibition, Bioletti had the creative task of attempting to come up with uses for the wine grape other than producing alcohol.

Bioletti retired in 1935 and died four years later in 1939.

PRESERVING AND CANNING FOOD: JAMS, JELLIES AND PICKLES

Food preservation has permeated every culture, at nearly every moment in history. To survive in an often hostile and confusing world, ancient man was forced to harness nature. In cold climates he froze foods on the ice, and in tropical areas, he dried them in the sun. Today, methods of preserving food commonly involve preventing the growth of bacteria, fungi (such as yeasts), and other micro-organisms, as well as retarding the oxidation of fats that cause rancidity. Many processes designed to conserve food will involve a number of different food preservation methods. Preserving fruit by turning it into jam, for example, involves boiling (to reduce the fruit's moisture content and to kill bacteria, yeasts, etc.), sugaring (to prevent their re-growth) and sealing within an airtight jar (to prevent recontamination). Preservation with the use of either honey or sugar was well known to the earliest cultures, and in ancient Greece, fruits kept in honey were common fare. Quince, mixed with honey, semi-dried and then packed tightly into jars was a particular speciality. This method was taken, and improved upon by the Romans, who *cooked* the quince and honey - producing a solidified texture which kept for much longer. These techniques have remained popular into the modern age, and especially during the high-tide of imperialism, when trading between Europe, India and the Orient was at its peak. This fervour for trade had two fold consequences; the need to preserve a variety of foods - hence we see more 'pickling', and the arrival of sugar cane in Europe. Preserving fruits, i.e. making jams and jellies became especially popular in Northern European countries, as without enough natural sunlight to dry food, this was a fail safe method to increase longevity. Jellies were actually most commonly used for savoury items; some foods, such as eels, naturally form a protein gel when cooked - and this dish became especially popular in the East End of London, where they were (and are) eaten with mashed potatoes. Pickling; the technique of preserving foods in vinegar (or other anti-microbial substances such as brine, alcohol or vegetable oil) also has a long history, again gaining precedence with the Romans, who made a concentrated fish pickle sauce called 'garum'. 'Ketchup' was originally an oriental fish brine which travelled the spice route to Europe (some time during the sixteenth century), and eventually to America, where sugar was finally added to it. The increase in trade with the sub-continent also meant that spices became a common-place item in European kitchens, and they were widely used in pickles to create new and exciting recipes. Soon chutneys, relishes, piccalillis, mustards, and ketchups were routine condiments. Amusingly, Worcester sauce was discovered from a forgotten barrel of special relish in the basement of the Lea and Perrins Chemist shop! As is evident, the story of food preservation, and specifically the modern usages of jams, jellies and pickles encompasses far more than just culinary history. Ancient civilisations, nineteenth century colonialism and accidental discoveries all played a part in creating this staple of our modern diet.

PICKLING RIPE AND GREEN OLIVES.

By FREDERIC T. BIOLETTI.

The continuous increase in the crop of olives in the State, due to the coming into bearing of young orchards during the past few years, has resulted in a growing demand for information on the subject of olive-growing, and especially on the treatment of the crop in oil-making and pickling. It was to meet this demand in a general way that Bulletin No. 123, "Olives," was issued in 1899. The methods of pickling recommended in that bulletin seem generally to have given satisfaction, but many complaints have been received of failure where these methods are said to have been followed. Most of the cases of failure, when more closely investigated, were found to be due, not to any defect in the method as given in the bulletin, but to some neglect to closely follow the recommendations given. Usually some important step had been omitted or some precaution neglected. The most common mistakes have been neglect of frequent changes of water, the use of impure water, the omission of a thorough disinfection of the pickling vats with boiling water, and failure to adapt the strengths of lye-and-salt solution to olives of various grades and degrees of ripeness. In some cases, however, where the method seems to have been followed faithfully, the pickles have failed to keep long enough to be marketed, especially when shipped to Eastern points, where they have been subjected to wide changes of temperature.

EXPERIMENTS WITH PICKLING RIPE OLIVES.

In order to throw light on the causes of these latter failures, and to attempt to find some solution of the difficulties, it was determined to undertake a series of experimental picklings. For this purpose, a quantity of ripe olives was kindly donated to the Experiment Station by Mr. John Rock, of the California Nursery Company, at Niles, Alameda County. The olives sent were particularly suitable for the purpose, as they consisted of several varieties and were nearly all dead-ripe, in fact, many of them were considerably over-ripe, and as they had been picked and shipped without any special precautions, they were considerably bruised when received. As the water at Berkeley available for the purpose of pickling was far from the purest, and was in fact contaminated to a considerable extent with organic matter and micro-organisms, the conditions were very favorable for the investigation of some of the main difficulties encountered in preserving olive pickles.

The olives were received on February 16, 1899, and were of the following varieties : Gordal, Uvaria, Columbella, Regalis, Rubra, Mission, Manzanillo, Nevadillo Blanco, forty pounds of each; Picholine d'Aix, Rouget, twenty pounds of each; Sevillano, ten pounds; Oliviere, four pounds; Salliern, Da Salere, Pigale, two pounds of each.

On account of the over-ripe, soft, and bruised condition of most of the fruit, it was determined to use a weak lye-solution, and in most cases to use salt from the beginning of the process. The lye-solution was made up by dissolving 1.9 oz. of lye in one gallon of boiled water, that is at the rate of 1.5 %. An analysis of the lye-solution showed it to contain 1.4% of pure lye, the difference of one-tenth of a per cent being due to impurities and to lack of strict accuracy in measuring the water. The measurement, however, was accurate enough for practical purposes.

The olives were sorted over before being pickled, in order to have them of fairly uniform size in each lot. This is very necessary for the best results, both on account of the difference of time needed to extract the bitterness of fruit of different sizes, and of the preference of the consumer for pickles of uniform appearance. The small olives were therefore rejected, but no attempt was made to eliminate over-ripe or bruised fruit. Any which showed signs of mold or decay were of course thrown out. As the olives were used very soon after picking, however, there were very few of the last kind. Earthenware jars were used for pickling, and the olives were kept submerged by means of floating wooden covers. The jars had a capacity of four gallons each, except the jar in which the Sevillano was treated, which was smaller, on account of the small amount of fruit available. Eight experiments were made, each with a different variety. A full account of the method pursued and of the course of the experiments is given in the following pages under the heading of the variety employed.

The questions which these experiments were devised to answer were: 1. Could large, over-ripe, bruised olives, which were otherwise in good condition, be used to produce wholesome marketable pickles ? 2. Could such pickles be preserved in good order for a reasonable length of time ? Incidentally, some information was obtained as to the relative suitability of the various varieties for the production of marketable ripe pickles.

1. Gordal. Large, very even in size, dark reddish purple, good condition, very ripe.

Feb. 17, 1899, 5 p. m.   Covered with a solution of 1.4% lye and 2% salt.
"   18, 9 a. m.   Rinsed and left in water four hours.
"   18, 1 p. m.   Put in 2% salt brine; olives rather soft.
"   19, 1 p. m.   Put in fresh 2% brine; olives firmer, but bitter.
"   21.   Put in 3% salt brine.
"   24.   Put in 4% salt brine.
"   28.   Put in 6% salt brine.
Mar. 3.   Put in 6% salt brine; olives still slightly bitter.
"   7.   Put in 8% salt brine; bitterness almost gone.
"   28.   Put in 12% salt brine and divided into four lots:
   A.   Placed in an earthenware jar, in which the pickles were kept submerged by means of a floating wooden cover.
   B.   Placed in a fruit-preserving jar and untreated.
   C.   Placed in a fruit-preserving jar and heated to 80° C. (176° Fahr.) once.
   D.   Placed in a fruit-preserving jar and heated to 80° C. (176° Fahr.) three times on successive days.
Oct. 30.   Olives in open jar (A) in perfect condition; better flavored than the Mission and of tenderer texture, though sufficiently firm; too light in color.
Nov. 1.   Olives in lot A divided into three lots, as follows:
   E.   Placed in a fruit-preserving jar and left untreated.
   F.   Placed in a fruit-preserving jar and heated to 80° C. (176° Fahr.).
   G.   Placed in a fruit-preserving jar and after pouring off the brine and filling up with water heated to 80° C. (176° Fahr.).
Jan. 5, 1900.   B and D of the first bottling showed mold on top, but all the others had kept perfectly.

Result in nearly 32 months.   Oct. 5, 1901.

Lot B (not heated) soft and spoiled.

Lot E (not heated) soft and, though not spoiled, was much inferior to the heated samples in texture and flavor. There was no mold nor scum on top of the liquid, but a slight acidity showed the effect of fungous or bacterial fermentation.

Lot D showed a little mold on top of the liquid, but the flavor of the olives was unaffected.

The other heated samples were all in good condition, of good flavor, and showed no scum or mold and no discoloration of the brine. Lot G was in just as good order as the others, and was superior in that it contained less salt. The specific gravity of the brine indicated only 5% of salt, owing doubtless to absorption by the fruit.

2. Picholine d'Aix.   In fair condition, very uneven in size, many small unripe olives, but the rest very large and fine, dark reddish-black. Many of the ripe olives were badly bruised.

Feb. 17, 1899, 5 p. m.   Covered with a solution of 1.4% lye and 2% salt.
"   18, 1 p. m.   Rinsed and put in 2% brine; fairly firm, but variable in color.
"   19, 2 p. m.   Put in 2% brine; very bitter, many blistered.
"   21.   Put in 2% brine; still very bitter.
"   24.   Put in 2% brine; still very bitter; very firm.
"   28.   Put in 2% brine; still bitter.
Mar. 3.   Put in 4% brine.

Mar. 7. Put in 6% brine.
" 16. Put in 8% brine.
" 25. Put in 12% brine; placed in an open earthenware jar, in which the olives were kept submerged by means of a floating wooden cover.
Oct. 30. Olives in good condition, firm and well flavored; many show blisters, but they have kept perfectly. The color is rather light, the stone large, and the skin tough.
Nov. 1. Divided into three lots and placed in fresh 12% brine in glass fruit-preserving jars, and treated as follows:
    A. Sealed without further treatment.
    B. Sealed and heated to 80° C. (176° Fahr.).
    C. The brine replaced with water, the jar sealed and heated to 80° C. (176° Fahr.).
Jan. 5, 1900. All the samples have kept perfectly.

Result in nearly 32 months. Oct. 2, 1901.

Lot A has a scum on top of the liquid, but the olives are firm and good, though somewhat inferior in flavor to lots B and C, on account of a slight acidity, due probably to a slight fermentation. B and C have kept perfectly and are of excellent flavor. C is the best, on account of the smaller amount of salt it contains. This variety is inferior in flavor and texture to the Manzanillo and the Gordal, but equal to the Mission. It is inferior to the Mission in color, but somewhat darker than the Gordal.

3. REGALIS. In fair condition, varying in size from medium to large, color varying from greenish-white to red and red-purple (similar to Columbella in color, but of larger size); shows a good deal of sooty mold.

Feb. 17, 1899, 5 P. M. Washed and put in a solution of 1.4% lye and 2% salt.
" 18, 1 P. M. Liquid which shows .25% lye, replaced with a 2% brine. The olives are firm, but generally light and very uneven in color.
" 19. Put in 2% brine. Color very varied.
" 21. Put in 3% brine. Bitterness nearly gone.
" 24. Put in 4% brine.
" 28. Put in 4% brine. A few olives slightly bitter still.
Mar. 3. Put in 6% brine.
" 7. Put in 8% brine.
" 16. Put in 12% brine.
" 28. Divided into four lots, as follows:
    A. Placed in open earthenware jar, in which the olives were kept submerged by means of a floating wooden cover.
    B. Placed in a glass fruit-preserving jar and sealed.
    C. Placed in a glass fruit-preserving jar, sealed and heated to 80° C. (176° Fahr.).
    D. Placed in a glass fruit-preserving jar, sealed and heated to 80° C. (176° Fahr.), three times on successive days.
Oct. 30. Olives in the open jar (A) in good condition, firm, and of good flavor. Very varied in color, from white to dark, almost black. Sample A was divided into three lots and treated as follows:
    E. Placed in glass fruit-preserving jar in 12% brine, and sealed.
    F. The same as E, but heated to 80° C. (176° Fahr.) after sealing.
    G. The same as F, but placed in water instead of brine.
Jan. 5, 1900. All the lots have kept well.

Result in nearly 32 months. Oct. 10, 1901.

Lot B (not heated) is soft, moldy on top, flavor quite spoiled.

Lots C and D have kept well and are of good flavor and texture, but are too light colored. D (heated three times) has a very dark-colored brine.

Lot E (not heated) shows a slight scum on top of the liquid, but the olives are uninjured except for a slight acidity due to the growth of molds or bacteria. The olives are yellow and even in color, the brine clear and colorless.

Lot F is in perfect condition and of excellent flavor, firm but tender, of an even dark gray color, but a little too salty. The brine is clear, but dark colored.

Lot G has kept just as well as F, and the smaller amount of salt makes it preferable. The color of both olives and liquid is lighter than that of F.

This variety is an excellent one for ripe pickles, with the exception of its lack of color.

4. MANZANILLO. Bruised and over-ripe, but otherwise in good condition, of good size, but not of such fine appearance or so even in size and quality as the Gordal.

Feb. 17, 1899, 5 p. m.   Put in a solution of 1.4% lye and 2% salt.
"   18, 1 p. m.   Rinsed and put in 2% salt brine. The color is a uniform deep black, with the exception of a few imperfectly ripe olives, which are a little green on one side; they are all of good flavor and fairly firm; still a little bitter.
"   19.   Put in 2% brine; the ripest fruit has lost its bitterness.
"   21.   Put in 2% brine; many olives are still very bitter.
"   24.   Put in 2% brine; many olives still bitter.
"   28.   Put in 2% brine; still bitter.
Mar. 3.   Put in 4% brine; bitterness nearly gone.
"   7.   Put in 8% brine.
"   16.   Put in 12% brine.
Oct. 30.   In fine condition; deep black and of excellent flavor; a little softer than the Mission.
"   31.   Divided into three lots, and treated as follows:
    A. Put in glass fruit-preserving jar and sealed (brine 12%).
    B. The same as A, but heated to 80° C. (176° Fahr.) after sealing.
    C. The same as B, but covered with a layer of paraffin before sealing and heating.
Jan. 5, 1900.   All the samples appear good and show no mold on top.

Result in nearly 32 months. Oct. 7, 1901.

Lot A. A heavy scum on top of the liquid; olives soft, acid, and slightly rancid, of poor quality.

Lot B. Better than A, but not so good as C.

Lot C. Has kept perfectly; of excellent flavor and texture and deep uniform black color.

The Manzanillo is the best pickle of the series, and is equaled in flavor and texture only by the Gordal, which is inferior in color.

5. MISSION. In good condition; firm and less over-ripe than the Manzanillo.

Feb. 17, 1899, 5 P. M. Put in a solution of 1.4% lye and 2% salt.
" 19, 1 P. M. Rinsed and put in 2% salt brine; olives all uniformly black and fairly firm.
" 21. Put in 2% brine.
" 24. Put in 2% brine; still bitter.
" 28. Put in 2% brine; still a little bitter.
Mar. 3. Put in 4% brine.
" 7. Put in 8% brine.
" 16. Put in 12% brine.
Oct. 30. In good condition; black, firm, and of good flavor; a little mold around the edge of the cover, but the olives unaffected.
" 31. Divided into three lots, and treated as follows:
A. Put in glass fruit-preserving jar and sealed (brine 12%).
B. The same as A, but heated to 80° C. (176° Fahr.) after sealing.
C. The same as B, but heated to 90° C. (194° Fahr.). The bottle cracked during the heating, and the sample was lost.
Jan. 5, 1900. Both lots A and B seem to have kept well.

Result in nearly 32 months. Oct. 8, 1901.

Lot A shows a scum on top of the liquid, and the olives have a faint acid taste, due to bacterial or fungous fermentation; the quality is fair but not nearly equal to B, and the color is lighter.

Lot B is of excellent quality and in perfect condition; firm but not tough, dark brownish-black; the flavor is clean and good, but not quite equal to that of the Gordal or the Manzanillo.

6. COLUMBELLA. In fair condition, some of the olives shriveled; of good size, averaging a little smaller than the Regalis.

Feb. 18, 1899, 8:30 A. M. Put in a solution of 1.4% lye.
" 18, 3 P. M. Rinsed and put in a solution containing 1.4% lye and 2% salt.
" 19, 12 M. Put in 2% brine; still bitter and taste of lye.
" 21. Put in 2% brine.
" 24. Put in 2% brine; taste of lye, but little bitterness.
" 28. Put in 4% brine; not bitter.
Mar. 3. Put in 6% brine.
" 7. Put in 8% brine.
" 16. Put in 12% brine.
" 28. Divided into four lots, as follows:
A. Placed in an open earthenware jar, in which the olives were kept submerged by means of a floating wooden cover.
B. Placed in glass fruit-preserving jar, and sealed.
C. The same as B, but heated to 80° C. (176° Fahr.) after sealing.
D. The same as C, but heated three times to 80° C. (176° Fahr.) after sealing.
Oct. 30. Lot A (in open jar) in good condition, firm, sound, and of good flavor and texture; color light and somewhat spotted.
" 31. Divided A into three lots, as follows:
E. Put in glass fruit-preserving jar, and sealed.
F. The same as E, except that it was heated to 90° C. (194° Fahr.) after sealing.
G. The same as F, except that it was covered with a layer of paraffin before sealing and heating.
Jan. 5, 1900. Lot B is covered with a thick mold; the others have kept perfectly.

Result in nearly 32 months. Oct. 10, 1901.

Lot B (unheated) moldy and spoiled.

Lots C and D (heated) firm, sound, and of good flavor.

Lot E (unheated) shows no scum on top and has kept fairly well; it has less of the acidity that characterizes most of the other unheated samples, but it is inferior in flavor to C, D, F, and G.

Lots F and G are firm, sound, and of good flavor, somewhat darker in color than E, which is light yellow; the brine is also darker in color than that of E.

7. ROUGET. In good condition, firm, ripe, but not over-ripe, uninjured by frost or rain; rather small, but larger than the Rubra, and nearly as large as the Nevadillo.

Feb. 18, 1899, 8:30 A. M. Put in 1.4% lye-solution.
"   18, 3 P. M. Put in 2% salt brine.
"   19, 12 M. Put in a fresh 2% salt brine. The olives have a sweetish taste and very little bitterness or taste of lye; the flesh is soft.
"   20. Put in 4% brine.
"   24. Put in 4% brine.
"   28. Put in 4% brine.
Mar. 3. Put in 6% brine.
"   8. Put in 8% brine.
"   16. Put in 12% brine; placed in open earthenware jar, in which the olives were kept submerged by means of a floating wooden cover.
Oct. 30. Perfectly sound, firm, and of a dark gray color. The flavor is not very good—too oily.
Nov. 1. Divided into three lots, and treated as follows:
          A. Put in a glass fruit-preserving jar, and sealed.
          B. The same as A, except that it was heated to 80° C. (176° Fahr.) after sealing.
          C. The same as B, except that it was covered with a layer of paraffin before sealing and heating.
Jan. 5, 1900. All the lots have kept well.

Result in nearly 32 months. Oct. 10, 1901. All the samples have kept well, except that A has a scum on top of the liquid. The heated samples are darker-colored and slightly better in flavor than A, but none of them are very good. The olives are tough and of poor flavor; the variety is evidently unsuited for pickling.

8. SEVILLANO. Extremely large and of fine appearance, but over-ripe and in poor condition. Many have been injured by being pecked by birds. The best were sorted out and pickled.

Feb. 17, 1899, 5 P. M. Put in a solution of 1.4% lye and 2% salt.
"   18, 9 A. M. Rinsed and put in 2% salt brine. The olives are of a uniform deep black, but very bitter and rather soft.
"   19. Put in 2% brine; still bitter, some soft but most are firm.
"   20. Put in 2% brine.
"   21. Put in 4% brine.
"   23. Put in 4% brine; still a little bitter.
"   28. Put in 6% brine; a very little bitterness left.

7

Mar. 3. Put in 8% brine.
" 7. Put in 12% brine.
" 16. Divided into three lots, and treated as follows:
    A. Put in a glass fruit-preserving jar and sealed.
    B. The same as A, but heated to 70° C. (158° Fahr.) for 15 min. after sealing.
    C. The same as B, but heated to 70° C. (158° Fahr.) for 15 min. three times on successive days.
Jan. 5, 1900. Lots A and B show mold on top of the liquid. Lot C has kept perfectly.

Result in nearly 32 months. Oct. 10, 1901.

Lot A quite spoiled.

Lot B shows a little mold on top, but the olives are firm and edible, though not of so good flavor as lot C.

Lot C has kept perfectly and the olives are firm and of very good flavor and dark color.

Over-ripe Olives.—These experiments show that even soft, over-ripe olives may be successfully pickled by proper modifications of the lye-and-salt method, even when the fruit has been somewhat carelessly handled before pickling and when the water used is not of the purest. The main precautions in such cases are to use a certain amount of salt from the beginning of the process, and to watch carefully for the first appearance of scum or slime on top of any of the liquids in which the olives are immersed. On the appearance of the slightest of these signs of fermentation, the solution must be changed and the receptacle thoroughly disinfected with *boiling* water. The salt hardens the flesh and makes it more resistant to fermentative organisms which exist in the water, and at the same time the antiseptic properties of the salt, even when used in such small proportions as 2%, are probably of use in delaying the increase of these organisms, molds, and bacteria. All the samples, with the exception of the Sevillano, kept without perceptible deterioration for eight months in open jars after pickling, although they were unprotected from the air except for a floating wooden cover. A ring of mold formed around the edge of the cover, but there was no perceptible injury to the flavor of the pickles, except for a slight moldiness in the taste of the Sevillano.

SUMMARY OF THE RESULTS OF VARIOUS METHODS OF KEEPING THE PICKLED OLIVES.

1. Seven samples (Gordal, Picholine, Regalis, Manzanillo, Mission, Columbella, and Rouget) were placed in 12% brine in open earthenware jars with floating wooden covers. A little mold formed around the edge of the floating cover, but the pickles were uninjured in flavor, texture, or color at the end of eight months.

2. Four samples (Gordal, Regalis, Columbella, and Sevillano) were placed in 12% brine in hermetically sealed fruit-preserving jars immediately after pickling. All except the Regalis were very moldy at the

end of eleven months, and they were all quite spoiled at the end of thirty-two months.

3. Seven samples (Gordal, Picholine, Regalis, Manzanillo, Mission, Columbella, and Rouget) were kept in open jars for eight months (see No. 1) and then placed in fresh 12% brine in sealed fruit-preserving jars. When examined at the end of eleven months (three months after sealing) they were all sound, but at the end of thirty-two months the Manzanillo was almost spoiled, the Gordal much deteriorated, and the others though still edible were much injured in flavor and appearance.

4. Three samples (Gordal, Regalis, and Columbella) were placed in 12% brine in sealed fruit-preserving jars and heated to 80° C. (176° Fahr.) immediately after pickling. They were all in perfect condition when examined at the end of thirty-two months. A sample of Sevillano, which was treated in the same way except that it was heated to only 70° C. (158° Fahr.), had deteriorated slightly but was still quite edible, while a corresponding unheated sample was quite spoiled.

5. Seven samples of the pickles kept in open jars for eight months (see No. 1) were placed in sealed fruit-preserving jars at the end of the eight months and heated to 80° C. (176° Fahr.). All the samples kept perfectly with the exception of the Manzanillo, which showed a slight deterioration in flavor.

6. Three samples (Gordal, Picholine, and Regalis) of the pickles kept in open jars (see No. 1) were placed in pure water in sealed fruit-preserving jars at the end of the eight months and heated to 80° C. (176° Fahr.). All kept perfectly and were preferable to all the others at the end of thirty-two months, on account of the smaller amount of salt that they contained. The specific gravity of the brine at the end of the thirty-two months indicated only 5% salt.

7. Three samples (Gordal, Regalis, and Columbella) were heated to 80° C. (176° Fahr.) three times on successive days in sealed fruit-preserving jars and kept perfectly, but showed no superiority to those heated only once. One sample (Sevillano) heated in the same way three times to 70° C. (158° Fahr.) also kept perfectly, and was superior at the end of thirty-two months to a sample of the same variety heated only once to 70° C. (158° Fahr.).

8. Three samples (Manzanillo, Columbella, and Rouget) were placed in sealed fruit-preserving jars and covered with a layer of paraffin before being heated to 80° C. (176° Fahr.). They all kept perfectly, but the paraffin was objectionable, and though the Manzanillo sample seemed to have kept a little better than the heated sample without paraffin, a little longer or higher heating would doubtless be equally effective, and the troublesome and unsightly paraffin would not be needed.

In a general way it may be said that all of the eleven *unheated* samples kept in fair to good condition for eleven months, but that they were all more or less spoiled before the end of thirty-two months; and, on the other hand, that all the twenty samples *heated to 80° C.* (176° Fahr.) kept perfectly to the end of thirty-two months, and were quite as good if not better than when first made, with the exception of one sample, which had deteriorated slightly.

It may be concluded from this that heating to 80° C. (176° Fahr.) is a sufficient means of preserving ripe olives, even in weak brine, for an indefinite period in hermetically sealed glass jars, provided that they are exposed to no greater changes of temperature than occur in an ordinary room in Berkeley. Whether they would have kept so long without deterioration if exposed to the trying conditions of shipment to Eastern points is doubtful; but there is no doubt that they would have kept much better and longer, even under those conditions, than olives treated in the usual way, and there is every reason to believe that a slightly higher heating, say to 90° or 95° C. (194° or 203° Fahr.), would have made them perfectly secure in any climate.

That this heating could be applied successfully to *olives in wooden casks* seems highly probable. A method which would doubtless be effective would be to place the pickles in a 8% or 10% brine in casks previously sterilized by steaming, and then to heat them with a current of pressure steam conducted directly into the brine. It would not be difficult to devise means for preventing the superheated steam from coming directly in contact with the olives and to keep them in constant movement in order to guard against too much heating of a part of the contents of the cask until they were all heated to a uniform temperature of 90° C. (194° Fahr.). Properly pickled olives are, moreover, very resistant to heat, and even 100° C. (212° Fahr.) continued for half an hour has no perceptible effect on their texture or flavor, unless it be a slight improvement of both.

The only objection to heating noted was that it causes a diffusion of the coloring matter of the olives into the brine, so that after heating the olives were lighter-colored and the brine darker than before. This diffusion, however, takes place in time even with unheated olives, and at the end of thirty-two months the unheated olives were in most cases actually lighter-colored than those which had been heated.

RELATIVE QUALITY OF THE VARIETIES TESTED.

All the varieties experimented with made pickles of excellent quality with the exception of the Rouget, which was small, greasy, and of a sweetish disagreeable flavor. Some of the varieties, however, were superior to others in flavor and texture, others in size and shape, and others in color. In the following lists they are arranged in the order of their merit in these various particulars:

Flavor and texture: Gordal, Manzanillo, Columbella, Regalis, Mission, Sevillano, Picholine.

Color: Manzanillo, Sevillano, Mission, Picholine, Gordal.

Size: Sevillano, Picholine, Manzanillo, Gordal, Mission, Regalis, Columbella.

In arranging these varieties in order of excellence, account must be taken of whether we mean excellence from the market point of view or from that of the connoisseur. The market demands two qualities above all others—large size and deep color. The typical olive shape is also desirable, but flavor and texture seem to be of secondary importance. For this reason no small olive will command a good price, and light-colored varieties such as the Columbella and Regalis, though of superior quality, will bring lower prices than the Mission or Picholine. The Manzanillo combines the qualities of large size, dark color, and delicate flavor and texture in a higher degree than any other variety; but its peculiar short, apple shape is considered less desirable by buyers than the typical olive shape of the Sevillano and Mission. For these reasons the relative excellence of the varieties tested would seem to be that indicated by their position in the following lists:

For home use: Gordal, Manzanillo, Columbella, Regalis.

For market: Sevillano, Mission, Picholine, Manzanillo, Gordal.

EXPERIMENTS IN PICKLING GREEN OLIVES.

At the suggestion of Mr. H. H. Moore, of Stockton, a series of experiments was undertaken with the object of determining the best method of pickling green olives. Green olives can be, and have been, pickled in California by exactly the same methods used for ripe olives, and when treated in this way leave nothing to be desired as regards flavor and keeping qualities. They have, however, the defect, fatal commercially, of losing their bright green color during the process of pickling, or shortly afterward. Practically all the unripe olives prepared in California turn an unsightly brown or gray, and have for this reason been unmarketable in competition with the imported Spanish olives, which usually retain their bright green or yellowish-green color, even when taken from the brine and exposed to the air for a considerable time on the counters of the grocers. In spite of the opinions of connoisseurs and of the analyses of chemists, which show that the ripe pickled olives are not only more pleasing to the cultivated palate but more digestible and nutritious, the fact remains that the favorite olives with the average consumer, and those for which the trade can afford to pay the highest price, are the large green "Queen" olives of Spain. These are the product of several large-fruited varieties pickled when green by processes which are trade secrets of the Spanish producers, or which having been devised for local conditions are inapplicable here. The finest and largest are made from a variety called Sevillano; though other large kinds are used, and doubtless any large olive such as Macrocarpa, True Picholine, Santa Catarina, and Ascolano, could be successfully marketed, if cured in the same way. Now that these varieties are beginning to be produced in California in notable quantities, it is important that some way of preparing them should be found that will enable them to hold their own against their imported rivals.

The olives used in the experiments detailed below were of the Agostino variety, and were kindly furnished by Mr. Moore from his orchard near Lodi, in San Joaquin County. They were shipped to Berkeley in ordinary wooden boxes, but arrived in good condition, for being hard and unripe they were not at all bruised. The greater part of them were quite green, but a few, 15% or 20%, had commenced to show indications of ripening in a tint of red on one side. Before using, they were divided into two lots of different sizes by means of an improvised grader consisting of a galvanized wire-netting screen.

As it had already been demonstrated that the best results with ripe olives, both as regards color, flavor, and keeping qualities, were to be obtained with weak lye-solutions applied several times, instead of a single treatment with a stronger solution, a first series of experiments was undertaken to demonstrate the effect of these two methods upon green olives. For this series nine earthenware jars of three gallons capacity were taken and filled with the olives to be experimented on. The jars had flat earthenware tops that were supported by a rim on the inside of the slightly constricted necks of the jars, so that when the various solutions were poured in and the jars filled, the weight of the tops kept all of the olives submerged. This complete submersion of all the olives is very important in obtaining uniform color. Each of the nine jars represented a different experiment, as follows:

Experiment No. 1. An attempt was made in this experiment to extract the bitterness without the use of lye. Three gallons of the large olives were placed in a jar and kept in a constantly running stream of water. The water entered at the bottom of the jar, flowed through the mass, and emerged at the top. At the end of six days the color was completely spoiled. The olives had become of a dirty brown, and had completely lost all their greenness, while the bitterness had hardly diminished at all.

Experiment No. 2. This was a duplicate of Experiment No. 1, with the exception that small olives were used. The result was identical.

These experiments showed among other things that a slowly-moving, constant stream of water is much more favorable to the growth of the bacterial and fungous slime which is so troublesome in pickling, than standing water that is removed even so seldom as once a day. The amount of slime produced in the six days of the experiment was sufficient to have spoiled the flavor of the pickles permanently, even had the color not been changed.

Experiment No. 3. Three gallons of large olives were placed in a $\frac{1}{2}$% solution of lye for twenty-four hours. The lye was then replaced with water which was changed every twelve hours. At the end of three days, the olives being still bitter, they were placed in a new lye-solution of the same strength for twenty-four hours. As the bitterness failed to disappear this lye-treatment was repeated at 6, 10, 13, and 16 days, being prolonged for only twelve hours, however, in these cases. Even with these repeated lye-treatments the bitterness had not all disappeared on the fifteenth day, and by that time the color was becoming brownish. On the twentieth day the bitterness had disappeared, but the color was spoiled.

Experiment No. 4. Three gallons of small olives were placed in a $\frac{1}{2}$% lye-solution for twenty-four hours; the lye then replaced with water, which was changed twice a day. In thirteen days the olives had commenced to turn brown and were still very bitter. At sixteen days they were placed in a $1\frac{1}{2}$% lye-solution for twelve hours, and at eighteen days nearly all the bitterness had disappeared, but the color was completely spoiled.

Experiment No. 5. Three gallons of small olives were placed in a 1% lye-solution for eighteen hours, and the water changed twice a day thereafter. At fourteen days the color was still good, but the bitterness had not disappeared, so another 1% lye-solution was poured on and left for twelve hours. This was repeated at seventeen days, and at eighteen days the bitterness was gone, but the color was spoiled.

Experiment No. 6. Three gallons of small olives were treated with a $1\frac{1}{2}$% lye-solution for eighteen hours, and then placed in water changed twice a day. The color commenced to change in four days, and in eleven days was quite spoiled and the bitterness still strong.

Experiment No. 7. Three gallons of large olives were treated for eight hours with a $1\frac{1}{2}$% lye-solution and then with two daily changes of water. The second day another eight-hour immersion in a lye-solution of the same strength was given. In four days the color was quite spoiled and the bitterness still persisted.

Experiment No. 8. Three gallons of small olives were placed in a solution containing $1\frac{1}{2}$% lye and $1\frac{1}{2}$% salt, which was replaced with water at the end of eighteen hours. The water was changed twice a day, but at the end of three days the color was completely spoiled.

Experiment No. 9. Three gallons of small olives were treated for four hours with a 2% lye-solution, and then placed in water changed twice a day. At the end of the second day the color had commenced to change, and at the end of three days it was completely spoiled.

These experiments showed that none of the usual methods of treating ripe olives, viz: short treatment with comparatively strong lye, long or repeated treatments with weak lye, or treatment with water alone, were successful in preserving the green color of unripe olives under the conditions of these experiments. By reference to the following table it will be seen that the green color was destroyed the more rapidly the stronger the lye-solution in which they were placed. In the jars where no lye was used the time of the destruction of the color was intermediate between that of the strong and that of the weak lye-solutions.

Nature of treatment.	Time of turning brown.
No lye—running water	6 days.
Lye ½%, repeated at intervals 4 times	20 "
Lye ½%, and after 16 days 1½%	18 "
Lye 1%, and after 14 days 1% again	18 "
Lye 1½%	11 "
Lye 1½%, and after 2 days 1½% again	4 "
Lye 1½% and salt 1½%	3 "
Lye 2%	3 "

In considering the experiments with lye in this table it would seem that the lye was the cause of the destruction of the green color; but the destruction of the color in those experiments where no lye was used indicated that this was not a full explanation. A jar of untreated olives was left in water accidentally for several days without changing, and it was noted that these retained their green color longer than the olives left in a running stream of water. This suggested that the turning brown was due to oxidation caused by the air dissolved in the running water and in the water replaced daily in the lye-treated olives; and that this oxidation was simply accelerated by the softening action of the lye upon the tissues of the fruit. Another series of experiments was therefore undertaken to see whether it was possible to preserve the green color by extracting the bitterness with a minimum exposure to the oxygen of the air or of that dissolved in water. This series was made with another lot of the same olives used in the first series. Lye-solutions of various strengths were used, and the olives kept in them until all the bitterness was removed. With the weaker solutions it was necessary to renew the solutions several times in order to completely neutralize the acrid substances to which the bitterness is due. All the lye-solutions were boiled before using, in order to expel all the dissolved air; earthenware covers were submerged in the brine.

SECOND SERIES OF EXPERIMENTS.

Experiment No. 1, with ¼% lye-solution:

Nov. 3, 9:45 P. M. Placed in lye-solution.
" 4, 9:45 A. M. Olives and solution unchanged in color.
" 4, 4:30 P. M. Lye not neutralized, uncolored.
" 5, 11:30 A. M. Lye nearly neutralized, uncolored.

Nov. 6, 11:30 A. M. Renewed lye-solution; olives green, but bitter.
" 8, 9:00 A. M. Renewed lye-solution; olives green, but bitter.
" 10, 8:30 A. M. Renewed lye-solution; olives green, but bitter.
" 12, 10:30 A. M. Renewed lye-solution; olives green, but bitter; lye straw-colored.
" 15. Olives still slightly bitter; put in 1% salt brine.
" 16. Olives still slightly bitter.
" 17. Put in 2% salt-solution; bitterness almost gone.
" 18. Put in 4% salt-solution.
" 20. Put in 6% salt-solution; olives turned a little brown.
" 22. Put in 12% salt-solution.

Experiment No. 2, with ½ % lye-solution:

Nov. 3, 9:45 A. M. Placed the olives in a ½% lye-solution.
" 4, 9:45 A. M. Lye and olives unchanged in color.
" 5, 11:30 A. M. Lye pale straw color; not neutralized.
" 8, 9:00 A. M. Olives still of good color, but bitter; lye neutralized; solution renewed.
" 10. Lye neutralized; renewed solution; color of olives darkening.
" 15. Put in 2% salt-solution; bitterness almost gone.
" 17. Put in 4% salt-solution.
" 20. Put in 6% salt-solution; olives a little brown.
" 22. Put in 12% salt-solution; olives a little browner.

Experiment No. 3, with ¾ % lye-solution:

Nov. 3, 9:45 P. M. Placed the olives in a ¾% lye-solution.
" 4, 9:45 A. M. Lye faintly yellow; olives unchanged.
" 6, 5:00 P. M. Lye neutralized; solution renewed; olives still bitter.
" 8, 9:30 A. M. Lye neutralized; solution renewed; olives still bitter.
" 9. Put in 2% salt-solution; bitterness disappeared.
" 12. Put in 4% salt-solution.
" 15. Put in 6% salt-solution.
" 20. Put in 8% salt-solution.
" 22. Put in 12% salt-solution.

Experiment No. 4, with 1 % lye-solution:

Nov. 3, 9:45 P. M. Placed the olives in a 1% lye-solution.
" 4, 9:45 A. M. Lye yellow; olives unchanged.
" 6, 11:30 A. M. Lye not neutralized; olives still bitter.
" 6, 5:00 P. M. Put in 2% salt-solution; lye neutralized; olives still a little bitter.
" 7. Salt-solution colorless.
" 8. Put in 4% salt-solution.
" 12. Put in 6% salt-solution.
" 15. Put in 8% salt-solution.
" 20. Put in 12% salt-solution.

Experiment No. 5, with 1½ % lye-solution:

Nov. 3, 9:45 P. M. Placed the olives in a 1½% lye-solution.
" 4, 9:45 A. M. Lye yellow; olives unchanged in color.
" 6, 11:30 A. M. Lye not neutralized; bitterness gone; color still good.
" 6, 11:30 A. M. Put in 2% salt-solution.
" 8. Put in 4% salt-solution; salt-solution colorless.
" 12. Put in 6% salt-solution.
" 15. Put in 8% salt-solution.
" 20. Put in 12% salt-solution.

Experiment No. 6, with 2% lye-solution:

Nov. 3, 9:45 p. m.   Placed the olives in a 2% lye-solution.
"   4, 9:45 a. m.   Lye brownish-yellow; olives unchanged in color.
"   5, 11:30 a. m.   No discoloration of the olives.
"   6, 11:30 a. m.   Lye not neutralized; bitterness gone; olives of good color and
like those of Experiment No. 5, except that they were a little soft.
"   6.   Put in a 2% salt-solution.
"   7.   Put in a 4% salt-solution; salt-solution yellow.
"   9.   Put in a 6% salt-solution.
"   12.   Put in an 8% salt-solution.
"   15.   Put in a 12% salt-solution.

Experiment No. 7:

A duplicate of Experiment No. 6, except that after placing in the 12% salt-solution the pickles were put into a glass vessel, sealed up and heated to 99° C. (210° Fahr.).

The pickles from each experiment were placed in bottles as soon as they were·in a 12% salt brine. At this time they were all of good flavor. Those of Experiments Nos. 1 and 2 ($\frac{1}{4}$% and $\frac{1}{2}$% lye) were slightly brown, but the color of the others was excellent. Those of Experiments Nos. 6 and 7 (2% lye) were slightly softened, and their flavor was perhaps not quite so good as that of those treated with weaker solutions of lye. All the pickles remained in bottles of clear glass exposed to the light for ten months, when they were examined with the following results:

After ten months.—Experiment No. 1. The olives were of very good flavor and texture and had kept perfectly, but were brownish in color, and the brine in which they had been kept was also dark brown.

Experiment No. 2. The olives were indistinguishable from those of Experiment No. 1, but the brine, though brown, was less dark-colored.

Experiment No. 3. The olives were of a golden-yellow color and equal to those of Experiments Nos. 1 and 2 in flavor and texture. Half of the brine had leaked out of the bottle, but the olives in the air above the brine had not suffered at all, and were equal in color and flavor to those still'immersed. The brine was brownish.

Experiment No. 4. The olives of this lot resembled those of Experiment No. 3, but were a little greener and brighter n color. The brine was a dark straw color.

Experiment No. 5. The olives of this lot were still greener than those of the foregoing experiment, and showed but little of the yellow color. They were about of the color of the imported "Queen" olives. The brine was colored only faintly yellow.

Experiment No. 6. The olives had kept perfectly their bright green color, and the brine was clear and almost colorless.

Experiment No. 7. The olives had the yellowish-green color of those of Experiment No. 5, but the brine was dark-colored, even darker than that of Experiment No. 1. The olives, both of this experiment and of the preceding, had become quite firm and had kept just as well as any, but the flavor was not quite equal to that of those pickled with the use of weaker lye-solutions.

Two months later the pickles were examined again, with results identical with those given above. Those of Experiment No. 3 which were exposed to the air had not suffered at all. The brightest green was that of Experiment No. 6, but the color that would be most acceptable to the trade would probably be that of Experiments Nos. 5 and 7. The objection to No. 7 would be the dark color of the brine. The slight yellowish tint of the brine of Experiment No. 5 would probably not be a serious objection, and the flavor of the olives was distinctly superior to that of Experiments Nos. 6 and 7. Supplementary experiments made later showed that when too strong a lye-solution was used the olives were somewhat bleached, and instead of the bright green of Experiment No. 6 or the yellowish-green of Experiment No. 5, we obtain a pale whitish-green, which is undesirable.

CONCLUSIONS.

This series of experiments shows that it is possible to produce green pickled olives which will retain their color for at least twelve months by the lye-and-salt method of treatment, if properly modified and controlled. The color is preserved, so that exposure to the air *after* the completion of the pickling process does not seriously affect the color for some time. (See Experiment No. 3.) The following process, based upon these experiments, is recommended:

Choice of Fruit.—Only large-fruited varieties should be used, as the small green pickles bring a very inferior price. The olives should be gathered as soon as they have reached full size and before they have colored notably. A slight pink color on one side does little harm, as it disappears during the process, but olives which have reached the stage of ripeness indicated by this first change of color will probably have less of the bright green than if gathered earlier. No two varieties should be pickled together, and the olives should be graded into three or four sizes. The reason for this is that different varieties and different sizes are almost sure to require different strengths of lye-solution, and it is therefore impossible to attain the best results unless this selection is made. The proper strength of lye-solution to use in each case is best determined by a preliminary trial, as follows:

Preliminary Trial.—Take a series—about six—of pint preserving-jars and fill them with the olives to be tested. Pour into them, respectively, a $\frac{1}{2}$%, 1%, 1$\frac{1}{2}$%, 2%, 2$\frac{1}{2}$%, and 3% lye-solution, sufficient to completely cover the fruit. At the end of forty-eight hours examine them. (It has been found that a sufficiently strong lye-solution will extract the acid and bitter principles of even very bitter olives in forty-eight hours.) At the end of this time some of the weaker lye-solutions will be found to have been neutralized, that is to say all the lye will have been used up in acting upon the acids of the fruit. This will be made evident by the lack of the slimy feeling which the fingers have when dipped into a lye-solution and rubbed together. Suppose that the $\frac{1}{2}$%, 1%, and 1$\frac{1}{2}$% solutions are neutralized, and that the 2% still has a slight slimy feeling. This will show that a 2% solution is a little stronger than is necessary to neutralize all the bitter or acrid matters in the sample tested. If, now, we use a 2% solution in curing the bulk of the olives from which the sample was taken, we are able to preserve the green color perfectly. If we use a somewhat stronger solution, say a 2$\frac{1}{2}$%, the color will bleach out a little; while if we use a much weaker solution, say a 1%, the green will change to that disagreeable gray or brown which we wish to avoid.

Process.—The appropriate strength of lye-solution having been determined, the olives are placed in convenient receptacles, where they can be treated with a minimum exposure to light and air. For this purpose fifty-gallon barrels with very large bungholes (four or five inches in diameter) and spigots are useful. After filling the barrels with olives the lye of the strength determined in the preliminary trial is poured in. Each barrel should be quite full of olives, and sufficient lye-solution should be put in to come flush with the bunghole. At the end of forty-eight hours the lye should be drawn off, the olives quickly washed with two changes of fresh water, and the barrels filled immediately with a 2% salt-solution. This brine should be replaced successively with a 4% and 8%, and finally a 12% solution, in the last of which the pickles remain permanently. The successive brines should be allowed to act for from forty-eight to seventy-two hours each, according to the size of the olives; the larger sizes requiring more time for the brine to penetrate and to displace the excess of lye which remains. The whole process will thus take from ten to fourteen days.

Absence of Air.—The essential part of the process is to avoid exposing the olives to the air during the pickling, until all the bitterness and acid are completely neutralized by the lye. After this the green color seems to be fixed, and exposure to the air does not change it much, though it is well, all through the process, to avoid leaving the olives uncovered by liquid any longer than necessary.

As different varieties of olives and even the same variety in different seasons and from different localities differ very much in bitterness, the importance of treating each variety separately is evident, as each will require lye-solutions of different strength to neutralize them. Very bitter olives, such as Mission, Sevillano, Manzanillo, and True Picholine, require solutions containing from $1\frac{1}{2}\%$ to $2\frac{1}{2}\%$ of pure potash lye, while olives containing little bitterness, such as Ascolano and Columbella, require only from $\frac{1}{2}\%$ to 1% solutions. As many of the commercial lyes are far from pure, some containing not more than 50% of potash, the number of preliminary tests must usually be at least six, as indicated above. Preliminary tests conducted as described do not require an analysis of the lye, though it is probable that lyes containing a large amount of common salt would act more slowly; and with such lyes a treatment exceeding forty-eight hours might be necessary.

To facilitate the preparation of the different strengths of solutions, it is convenient to remember that as a gallon of water weighs 128 ounces, one and a quarter ounce of solid lye is equal (in round numbers) to one per cent; or that one pound of such lye will make nearly twelve and a half gallons of one per cent solution.

Lightning Source UK Ltd.
Milton Keynes UK
UKHW010026291119
354412UK00003B/864/P